Everything Is Now

Everything Is Now

Revolutionary Ideas from String Theory

Bill Spence

CRC Press
Taylor & Francis Group
Boca Raton London New York

CRC Press is an imprint of the
Taylor & Francis Group, an **Informa** business

First edition published 2021
by CRC Press
6000 Broken Sound Parkway NW, Suite 300, Boca Raton, FL 33487-2742

and by CRC Press
2 Park Square, Milton Park, Abingdon, Oxon, OX14 4RN

© 2021 Taylor & Francis Group, LLC
CRC Press is an imprint of Taylor & Francis Group, LLC

ISBN: 978-0-367-49022-5 (pbk)
ISBN: 978-1-003-04403-1 (ebk)

Typeset in Minion
by Cenveo® Publisher Services

This unusual book provides a concise description of some of the most exciting developments in theories of fundamental particles and their forces that have emerged from the study of string theory. The underlying ideas involve a mix of subtle physical insights and sophisticated mathematics, which are difficult to convey to a non-specialist audience, However, in eleven short chapters Bill Spence manages to describe the essence of the subject with great clarity.

The subject of String Theory evolved from the assumption that the many different fundamental sub-atomic particles are identified with different kinds of vibration of an extended string-like object. It subsequently developed into a much broader subject that provides a mathematical framework for unifying the fundamental particles with the quantum geometry of space and time through which the particles move.

A striking feature of many developments in this area over the past forty years is the remarkable symbiosis of theoretical physics with modern mathematics. Not only has new mathematics transformed the understanding of string theory and quantum field theory but the developments in these areas of theoretical physics have fed into some of the most important developments in mathematics. This book succeeds remarkably well in conveying these key developments and, rather amazingly for a subject that is so closely connected to mathematics, it contains virtually no equations!

The book's main themes include seemingly esoteric subjects such as topological features of extra dimensions of space, the geometry of twistors, mirror symmetry and the "holographic" unification of general relativity (Einstein's theory of gravity) with theories of the other forces. All these are described in a manner that avoids hyperbole and over-simplification while avoiding all mathematical detail.

This short book will appeal to intelligent non-experts with no mathematical background who would like a brief overview of the field. It also provides a stimulating introduction for young people thinking of pursuing fundamental areas of physics or mathematics in more depth.

Professor Michael B Green FRS
Professor of Theoretical Physics
University of Cambridge/Queen Mary University of London
Pioneer of String Theory research

To Penny and Grace, for Everything

Contents

Acknowledgements

It is a pleasure to thank my past and current collaborators in research in string theory and colleagues in the Centre for Research in String Theory at Queen Mary University of London, in particular Andi Brandhuber and Gabriele Travaglini, with whom I have shared explorations of the extraordinary worlds opened up by the invention of twistor string theory in 2003. Thanks to Nathan Day, Penny Green, and Gabriele Travaglini for comments on the manuscript, and Grace Spence Green for design advice. Any errors or omissions are my own.

Author Bio

Bill Spence is Professor of Theoretical Physics and the founding Director of the Centre for Research in String Theory at Queen Mary University of London. www.billspence.org

CHAPTER 1

Introduction

THIS BOOK DESCRIBES THE revolutionary ideas emerging from string theory. These ideas are new, intensely strange, but utterly beguiling. Indeed, the entire history of physics can be seen as revealing strange properties of nature in radical ways that seem to contradict everyday experience. An early example of this was the discovery that the earth is not at the centre of the solar system. And around a hundred years ago, two profoundly radical and counter-intuitive theories totally revolutionized physics – relativity and quantum mechanics. Perhaps we are entering the time when string theory will similarly overturn our current understanding of the universe.

Whilst Einstein's special relativity showed that space and time were inextricably linked, it was his general theory that revealed the almost unimaginable idea that space and time, rather than being 'where and when things happen' could in themselves be tangible physics – that the actual force of gravity is a manifestation of the curvature of space-time. A little later the mysterious world of quantum mechanics was revealed – matter is neither a wave nor a particle, but somehow something intermediate which could shape-shift into either, depending on how you looked at it.

The last hundred years has seen intense work trying to take Einstein's work further, on two fronts. The first has been trying to prove that the other forces of nature, nuclear and electromagnetic, are manifestations of some property of space-time in the same way gravity is. The second has been trying to understand how the ideas of quantum mechanics could be applied to gravity, to space-time itself. There has been wonderful progress in understanding the physical world in this period. Special relativity has been successfully combined with quantum mechanics to create quantum field theory, and then this was used to formulate the so-called Standard Model, which describes all the non-gravitational forces that have been discovered. This theory has now been successfully tested to tremendously high accuracy and precision.

The entire history of physics has been a great process of unification, whereby previously apparently different or complex phenomena have been shown to be described much more simply as different manifestations of a more fundamental underlying theory. For example, the existence of all the chemical elements like hydrogen, iron, uranium, etc., can be understood as the different ways that just three types of particle (protons, neutrons, and electrons) can combine under nuclear and electromagnetic forces. The Standard Model in turn understands how these three types of particle, together with many others that have been discovered and which are related to them, can be understood in terms of a combined theory of just three forces.

How the curved space-time description of gravity fits with the other forces has been the outstanding question in fundamental physics for more than a century. But attempts to unify gravity and quantum theory have so far been epic failures. Many of us in theoretical physics feel that this is simply because we haven't been sufficiently radical in our ideas, and that string theory, whatever its eventual fate, will liberate our imaginations enough to light a path to future unification. The ideas emerging from string theory now may just be sufficiently crazy to achieve this liberation.

First, a few notes on the book. There is a wide gulf between the abstruse mathematical language of theoretical physics and the everyday written word; in attempting to bridge this I have inevitably taken much license. I have covered the ideas via a roughly historical route, but have not reviewed all the major areas of research in string theory. The subject has become tremendously broad and deep and this short book is intended to be simply a taster of the subject. There are other popular and more detailed works listed at the end of the book where further material can be found, although the subject moves extremely fast and the very latest material may not be covered.

You will find very few formulae here, as the book is intended for readers without formal mathematical training. But just to be clear, the properties of the theories described in this book can all be derived from the mathematics used to set them up. It is not the case that you can just say anything. In physics, once you have written down the theory, the consequences flow from the mathematical formalism and basic physical principles. This is what is so extraordinary about string theory; it is full of astonishing possibilities. This doesn't mean it is right, as the final judge is reality itself, accessed by repeatable experimental tests.

In summary, there are no qualifications needed for reading this book, no degrees or special scientific knowledge. You only need curiosity about the world and a readiness to stretch your imaginations past the limits of experience.

CHAPTER 2

Everything Is
Now – Then

THE UNITED KINGDOM IS fortunate to be the home of one of the world's most extraordinarily creative mathematicians – Roger Penrose in Oxford, who started producing seminal ideas in his youth and, currently past the age of 88, continues to do so. He has written some blockbuster (literally, some might double as house-bricks) books which attempt to describe his theories and more speculative proposals.

One of his early mathematical inventions led to a radical transformation of how we view space and time. This is his theory of twistors, originating in the 1960s. Prior to this, physicists described the world purely in terms of events happening at some place in space at some time, in other words at a 'point in space-time'.* This is hardly remarkable, but it is so obvious an assumption that it is, first, very hard to even recognize it as such and, second, even harder to imagine a different way to think.

Whilst quantum mechanics has been very successfully combined with special relativity to form quantum field theory, this has the rather unsettling feature that most of the calculations in the theory yield the same answer – infinity! (Or sometimes minus infinity, which is not much consolation.) This 'infinity' is not a number in itself, but a formal mathematical object that is bigger than any number you can think of, like what you would get if you kept adding $1 + 1 + 1 + \ldots$ forever.

---

* Think of 'space-time' as all the places that events can occur, combined with all the possible times that these events can happen; a 'point' in space-time is then an 'event' – a place and a time.

It is a bit much to ask our experimental colleagues to test this, as you always get some particular, finite number when you do any measurement. Fortunately, there is a long-standing procedure which is applied here, which puts another set of infinities with opposite signs into the mathematics at the beginning. These are used to cancel off the rest of the infinities coming from the original calculations. This has the reassuringly innocuous name of 'renormalization'. It is not the sort of thing any scientist feels comfortable with, and would never have been countenanced, except that it has two utterly convincing features. First, it is in fact a well-defined procedure, and second, it gives definite, finite answers that can be tested, in some cases agreeing with experiments with a precision of one part in a hundred million.

Some of the infinities that arise when you combine quantum mechanics and special relativity are due to events such as the interaction of particles with each other at very precise points in space-time. Forcing quantum particles to be at precise locations is known to cause various issues, as quantum objects like to be 'uncertain', so this is not surprising perhaps. One idea to deal with the infinities of quantum field theory is that if one could formulate it in a language that is not 'point-like', then one might resolve this problem. But this means finding a radically different conceptual approach, and the necessary mathematics to represent it.

This is indeed what happens in string theory, as we will see later, but first let us also think about what seems like a totally different question at first sight, the question of what 'mass' is. Some of the fundamental particles that make up all matter have mass, like the proton or electron, whilst others are massless, like the photons that light is made from. This is really a bit odd – what is this thing called 'mass' and why do some particles have it and some do not? Contrary to what common sense might tell you, there are reasons to think that 'mass' is not a fundamental property, but that it emerges as particles move. Intuitively, mass is a sort of resistance to being moved, and there is an analogue of this for massive quantum particles. This is that their mass might be due to being immersed in a sea of other quantum particles that drag upon them and slow them down.

How does this work? The basic reason is because in quantum theory there is really no such thing as 'nothing'. You might think simply of creating nothing, so to speak, by having a box, and taking everything out of it, so that what is left inside is 'nothing'. But what about the air that's still there? In physics, the inside of an empty box without any air or other matter inside is called a 'vacuum'. The space between the stars is pretty close to a vacuum – there are about a million molecules in each cubic centimetre (sugar cube sized region). It is actually extremely hard to take everything out of a box – laboratory vacuum chambers still contain ten billion molecules in that volume. Most of these molecules are hydrogen and helium. Compare this to the air we breathe, which has more like ten billion, billion molecules per cubic centimetre. Still, you might imagine that somehow you could catch all the molecules in a box and take them out, leaving nothing inside.

But this turns out to be impossible, as in quantum theory there is the counter-intuitive result that particles can be created out of nothing, as long as they disappear again quickly enough, roughly speaking. This can be understood as being due to an Uncertainty Principle. The commonly known Uncertainty Principle states that you can't know, at the same time, the exact position and momentum (mass times speed) of a quantum particle, and that there is an inverse relationship between these. The more accurately you know one, the less precision you can have with measurements of the other.

There is another Uncertainty Principle that relates energy and time in the same way, stating that the more you pin some event down in terms of when it happened, the less you know about the energy involved, and vice versa. Then the thing about the vacuum, due to this Uncertainty Principle, is that you can have some energy appearing in the form of particles, as long as this doesn't last long. Thus, the 'vacuum' of quantum theory, where all the real particles are taken out, is not then nothing, but is a sea of virtual particles which pop out of nothing and quickly disappear again.

Finally, then, what happens when a real particle is introduced into this vacuum is that it can bump into some of these virtual particles if they pop up nearby, thereby being knocked off course and slowed down, in everyday language. This does not necessarily create a mass for the real particle, but in some cases it does – for the Higgs particle, for example. This particle was predicted in 1964 and finally discovered in 2012, and behaves in just such a way – with a sea of virtual particles in the vacuum which acts to create mass for it.

Some of these ideas will resurface later, but the conclusion here is that we might consider mass as a derived quantity, a side-effect of something else, so that massless particles are somehow more elemental and funda-mental. Thus, let's concentrate on them.

The extraordinary feature of massless particles is that when they are free to move, they always travel at the speed of light. This is of course incredibly fast, which is why you can communicate via electronic means almost instantaneously with someone on the other side of the world. This is because the electric currents that encode voice, text, or video are caused by the interactions between charges, which are transmitted by photons and which happen at the speed of light. The speed of light is also a maxi-mum speed – if you try to push massive particles up to that speed, they start to become heavier and heavier (another prediction of special relativ-ity), so it gets harder and harder to accelerate them. You will never get the particle to reach the speed of light as this would require an infinite amount of energy.

The speed of light is really fast, but that's not actually what makes it most interesting. The next step is to think about another consequence of special relativity, which is that time slows down for moving objects. An observer sees a moving clock tick more slowly than a stationary one. We don't notice this, as our everyday world is full of relatively very slow objects in comparison, but it kicks in as speeds approach that of light. And, mathematically, at the speed of light clocks appear to slow down and stop completely. Whilst our time goes on and on, the clock on the photon doesn't move. In simple terms, as far as we are con-cerned, for a free massless particle Everything Is Now. Whilst this is a consequence of special relativity, we will see later that in an approach emerging from string theory, the fundamental quantum properties of interest are all made from real free massless particles. Hence the title of this book.

As discussed in the beginning of this chapter, Penrose suggested that instead of thinking about space-time as the basic arena where things hap-pen in physics, one should consider something more relevant to massless particles, as these are likely to be more fundamental. Massless particles travel at the speed of light, becoming like light rays. Thus, Penrose pro-posed to use the whole space of possible light-like rays as a new math-ematical underpinning for physics. This space is called twistor space, and one can think of twistors as like light beams. The idea is to use light beams instead of space-time points in the theory, with the hope that this would solve the problems of quantum field theory, and remove the infinities that it generates.

How can we picture this? It is relatively easy to visualize what 'space' is in the mathematical sense. For example, flat two-dimensional space is like the surface of a sheet of paper which is infinitely large. It is two-dimensional as we need to give two numbers to describe where any particular point is on the sheet – how far down the paper vertically and how far along it horizontally. Three-dimensional flat space is what we think of as all around us, and to describe where something is we need to say how far it is from where we are – forwards or backwards, left or right, and up or down. Four-dimensional space is harder to picture as you need to think of another dimension when we are aware of only three and there seems to be 'nowhere to put it'. Nevertheless, it can be defined mathematically by using four numbers for each point in the space. Four-dimensional space-time, with three space and one time dimension, can be thought of as the set of all 'events' – which can be described by where they happen (using three numbers) and when (using one number, the time of the event).

Visualizing twistor space is more difficult. A light ray, the beam from a torch, for example, seems simple, just a straight line in our three-dimensional space. In four-dimensional space-time, the fact that light always travels in empty space at the same speed means that it is represented mathematically by a line with a fixed slope. Because of this, an individual light ray emerging from some fixed point could be represented by any point on this line. Then, the space of light rays that could emerge from a given point, which looks like a sort of cone in space-time, can be identified with a set of individual points on that cone, which mathematically is in fact just the surface of an ordinary sphere. It is this sphere that lives in twistor space. In this picture, a point in space-time corresponds to a sphere in twistor space; and conversely, a point in twistor space corresponds to a light ray in space-time. The exact story is a little more complex (literally, using complex numbers), and there are various relevant pictures you can draw to illustrate a 'twistor'. One of these is shown in Figure 2.1.

Twistor space in generality is a four-dimensional complex space, which can also be thought of as describing pairs of spinors. These are objects that we will meet later. Twistor space is a beautiful mathematical construction with lots of interesting features that were followed up by a group of enthusiasts in the 1960s and later. But this approach just didn't catch on in physics. Whilst the basic idea was clear and people were sympathetic to it, in practice this didn't seem to provide a better way to work things out,

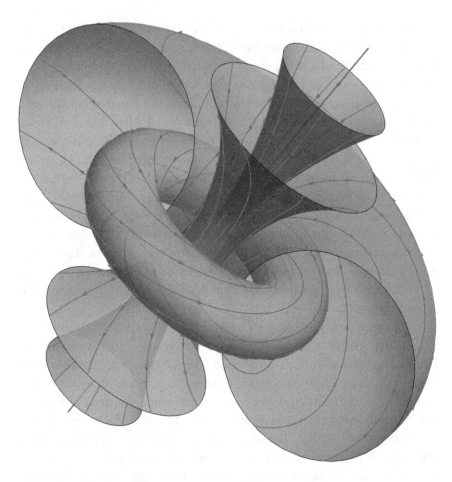

FIGURE 2.1 A twistor picture

or a more helpful conceptual framework for generating new progress. This potential trove of treasure was to remain essentially undisturbed for the next 40 years until, as we will see, string theory rediscovered and revitalized it for a new generation, enabling the rapid rise of entirely new ways of looking at quantum physics, and much associated progress.

CHAPTER 3

It's Right Behind You

A FAMILIAR TECHNIQUE OF CINEMA is where the subject of the camera is oblivious to some strange and probably threatening figure that emerges close behind them. You want to shout 'Look out, it's right behind you!' A similar idea is used in some films where a person appears to be living in parallel, unsuspected, inside false walls in the house of an intended victim.

A recurrent theme of physics is exactly that we are totally unaware of real things that are happening all around us. Luckily, for example, you do not feel the many billions of neutrinos – one of nature's odd little particles – that whizz through your body and then the entire earth, every second. You need an underground bath of a thousand tonnes of heavy water to catch these. There is also the still mysterious and unexplained fact that the galaxies that we observe behave as if they have much more in them than the stars that can be seen. In particular, for rotating galaxies like our own Milky Way, the rotation speeds of stars indicate that there is much more matter in the galaxy than is contained in the objects that can be seen. This has been named 'dark matter' for obvious reasons. It is probably all around us, and throughout the universe, but we don't know yet what this is made from. There is also dark energy, but that's another story. The two seem to account for about 95% of the mass in the universe, and this is a pretty good example of the fact that there is still plenty of stuff we don't know.

Dark matter and energy are unseen, but let us consider a more extreme idea of an unseen world – is it possible that the parallel world idea explored in film is true of space and time itself – that all around us there is another space that we cannot immediately see or feel, but which might help explain some of the physics that we don't understand. It seems at first that it would be hard to imagine how this would work. But it is not a new idea, and soon after Einstein formulated his Theory of General Relativity it was proposed by Kaluza and Klein that there might be a fourth space dimension, adding to the three that we are aware of, which might explain where the electric and magnetic forces come from. This proposal had the intriguing feature that this extra dimension was not 'spread out' like the space we inhabit. Instead, this extra dimension was rolled up everywhere into a tiny circle, so small that whilst we could not see it, the particles of physics could move in it.

You can think of everyday examples of tiny rolled-up dimensions. For example, from a distance a thin wire appears one-dimensional and it appears that an ant can only travel up and down the wire. See the picture on the left in Figure 3.1. But if you go closer, as in the picture on the right, you can see clearly that a small ant could also travel in the perpendicular direction around the wire itself, and that if that ant set off on this journey it would come back to where it had started after a while – it is going around in a small circle.

Now, if you imagine a space where the direction along the wire is replaced and expanded to be our three-dimensional space, and the direction around the wire is a rolled-up fourth space dimension, then you have the idea behind the Kaluza-Klein theory. For the ant on a wire in Figure 3.1, you can see that for every point in the one-dimensional space along the wire, there is also a tiny rolled-up dimension – the circle that goes around the surface of the wire itself at that point. Similarly, in the Kaluza-Klein

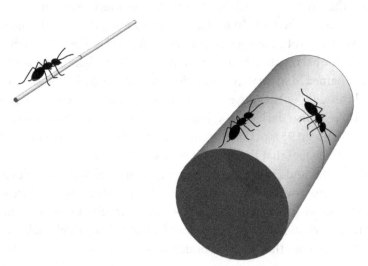

FIGURE 3.1   Rolled-up dimensions

theory, for every point in our three-dimensional space, there is a tiny rolled-up circle. At first sight, this theory seemed interesting, as it gives equations similar to those describing the electric and magnetic forces. In the end the Kaluza-Klein theory didn't work, but the idea behind it has been a powerful driver of research in string theory in the past 30 years. To understand why, we need to think a bit more about dimensions.

We experience three spatial dimensions all the time. We are free to move in three independent ways: up/down, left/right, and forward/backwards. Any other motion can be described as a combination of those three types of moves. That we have these three spatial dimensions is another feature of our world that is so obvious it is hard to imagine anything different. But the mathematics used in the theories of physics can often easily be adjusted to accommodate more than three dimensions, without anything seeming to go wrong. Thus, it is really a mystery from the point of view of physics as to why we live in three space dimensions. On the other hand, having more than one time dimension is much harder to accommodate, as it leads to all sorts of paradoxes.

When string theory emerged onto the scene in the 1960s, it released three extraordinary bombshells. One of these we will save for the next chapter. The other two are equally astonishing. To understand them, we first need to describe what this early string theory was. Prior to string theory, physics described the fundamental particles of nature as point-like objects, moving around and bumping into one another ('scattering'). However, the properties of some of these particles suggested that they had a string-like substructure, as they behaved as if they were made of two particles held together by some sort of stretchy string or glue. To explore this, a mathematical theory of moving and vibrating strings was devised.

When this string theory was investigated in detail, the utterly unexpected and bizarre fact was discovered that the theory only made sense if the space that the strings moved in had a particular number of dimensions. And this number was 25; or 26 if you include time. Not obviously very convenient for understanding our world it seemed, and most physicists paid little attention. Another reason to ignore the theory was that it predicted a massless particle that had not been found, one with quantum 'spin' equal to two in suitable units. To compound matters, it had a further particle, a 'tachyon' or particle that travels faster than light, something that is extremely hard to accommodate in any of our theories.

Whilst most people ignored this apparently crazy theory, a few wondered about this predicted massless particle with spin two in string theory.

Gravity, when considered from the point of view of particles and forces, predicts a massless particle with spin two which communicates the force, like the photon of light communicates the electric and magnetic forces. For gravity, this putative particle was called the graviton. The recent discovery of gravitational waves is strong evidence for the existence of the graviton.

Could string theory with its massless spin two particle be related to gravity in some way? This is where the second bombshell came in, a little later. Strings, it seemed, could only exist in 25 space dimensions. But what sort of space was this? Could it be curved, for example? This took a while to work out, but what emerged was the following: strings could only exist in a space-time that satisfied Einstein's equations of General Relativity!* In simple terms, not only could string theory dictate how many dimensions were allowed to exist, it knew that gravity had to exist as well. Whilst there is much more to this story, there are two key ideas to describe.

One is supersymmetry. We have talked about 'particles' and 'forces' as two different things. There are matter particles (like electrons) and force particles (like photons) which carry the forces between the matter particles. Roughly speaking, physicists have two names for these, fermions (matter) and bosons (force). But why should the particle world be divided thus? How does a particle know if it's meant to be matter or force? Physicists had found since the 1960s that matter particles form into family groups, with relationships amongst them called 'symmetries'. In a well-defined mathematical sense, two apparently different particles are two different ways of viewing the same object. Force particles also fell into families, albeit in a different way. Then, in the 1970s the idea was proposed that this might also be true between the matter and force particles, that they might be siblings, related by what was called 'supersymmetry'.

The theories that had this new feature of supersymmetry proved to be extremely interesting. For example, when you worked out the quantum predictions of the supersymmetric theories you didn't get infinity as the (nonsensical) answer quite so often, so you didn't have to do so much of the 'renormalization' mentioned earlier to get rid of these infinities. This and other related features caused a great deal of interest amongst researchers, and there has been an enormous amount of work done on supersymmetric theories in the past 40 years, some of which will form part of our later story. But just to make one comment here, the supersymmetric string

---

* Actually a modification of them in general.

theories required not 25 space dimensions like ordinary string theory, but a new number – nine.

The second key idea we need to describe comes back to the problem of what to do with the extra dimensions that string theory seems to need, and that we don't appear to have around us. We can roll these up into tiny circles, or something more complicated but still tiny, so that they are not obvious, and this works if the rolled-up space is a particular sort of geometry. These 'internal' geometries, required for the consistency of string theory, have some beautiful properties, and many of them have been studied by mathematicians, such as the 'Calabi-Yau' surfaces. The identification and analysis of allowed string geometries has been highly active for more than two decades. One of the most remarkable discoveries has been that of 'mirror symmetry', discussed in Chapter 4. This has inspired much subsequent work in pure mathematics.

Hence there is now considerable understanding of how small, rolled-up dimensions could occur in string theory, and explain why we don't see all the dimensions predicted by string theory. But could instead we have 'large' extra dimensions which are nonetheless not obvious? This would be like the films mentioned above, with the unseen parallel universe hiding behind some false walls of our own world. The answer to this question, found more recently, is indeed yes, that this is possible. The key requirement is that anything in the large extra dimensions needs to interact with our world only weakly, so that these dimensions won't be obvious to us, even though they are all around us. Of course, there does need to be interaction with us to some extent; otherwise, the extra dimensions would be truly invisible and lead to no predictions that we could eventually test.

This interaction does take place in particular cases in an interesting way. One of the features of some theories which have large extra dimensions is that the non-gravitational forces are less affected by the extra dimensions, but that the force of gravity applies throughout all the dimensions, and this dilutes it, so that here in three spatial dimensions gravity is a weak force compared with the others. This relative weakness is a mysterious feature of gravity, and is an obstruction to trying to combine gravity with the other forces of nature into one theory. It is easy to give an example of the relative weakness of the force of gravity: a small magnet can cause a paper clip to jump up off the floor when held near to it. The magnetic force from the small magnet exerts a greater attractive force on the clip than the entire six thousand billion, billion tonnes of the earth pulling the clip downwards by gravity.

CHAPTER 4

My Donut Has No Hole

As described in the abstract, string theory was found to dictate two fundamental properties of the space-time that it lives in. These are how many dimensions it has, and that it has to obey the equations of gravity, that is, of General Relativity. It's almost as if you can think of the strings, which can be loops or segments, as being able to explore the space they are in since they are extended objects, unlike point particles that can only be in one place at a time.

A third and stranger feature of string theory was also discovered when people explored how strings might behave in a space-time where some dimensions were rolled up. Imagine if one of the dimensions is rolled up into a tiny circle. If you consider a 'closed string' (i.e., a loop), then it can move around anywhere in space-time, including over the rolled-up circle. But it could do something else – it could wrap itself around the circle. How could this occur? The ends of an 'open string' (i.e., a segment of string) are allowed to join up to form a closed string, so if this happened when the open string was wrapping nearly all around the rolled-up circle, then you would get a closed string wrapping itself around the circle.

The size (radius) of the circle comes into the calculations describing this, and as you might expect, for circles of different sizes you get different results. But it turns out that the string theory for a circle of a given radius is the same as for a circle with the inverse value of the radius. As far as the strings are concerned, a circle of tiny radius is the same as one of enormous radius. This suggests that there might be a minimum as well a maximum distance in string theory. After all, as far as the strings are concerned, living on a tiny circle is the same as living on a giant one. What the strings think of as 'distance' seems to be something dramatically different to what we experience.

But it doesn't stop there. There is a whole field of mathematics called topology, which investigates when it is possible to continuously deform one thing into another, without breaking it in any way. This is easily understood for everyday objects. You could squash and stretch a soccer ball so that it had exactly the shape of a rugby ball, for example. On the other hand, if you imagine a donut without a hole (i.e., shaped like a somewhat squashed ball) then you can't deform it into a normal donut with a hole without breaking through its surface. This is easy to think about for such two-dimensional surfaces, as the surface of a torus and that of a sphere are really different since one has a hole through it, and in fact two-dimensional surfaces can be classified according to topology simply by how many holes they have. For higher dimensional spaces this understanding is much more difficult, and in some cases not yet achieved.

What if we think about the topology of the spaces that the strings are moving in? We just saw that the strings can 'explore' space by wrapping around circles. They can similarly also wrap around holes – just draw a loop on your donut that goes around the hole (see Figure 4.1). It turns out

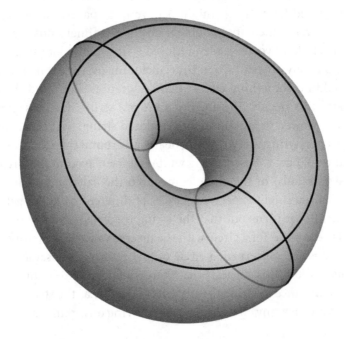

FIGURE 4.1 Strings winding on a torus

that when the theories of strings moving in different spaces are studied, then sometimes the string theories are exactly the same even when the spaces they are moving in are different. The strings don't 'see' some of the holes in the space.

This means that the whole theory of topology and geometry that has been derived to understand when spaces are equivalent needs to be redeveloped for string theory. We need a 'stringy' topology and geometry. We don't yet know precisely how to describe this but a lot of work is going on to find this out. To give some idea of this, we first need to correct a simplification in the discussions above, where we introduced just two types of strings. These were open strings, which have ends, and closed strings, which are loops. The ends of an open string can join to turn it into a closed string, and conversely a closed string can break and turn into an open string. This is still true, but when you put supersymmetry into the mix, so considering what are called 'superstrings', then you find, surprisingly, that there are five different varieties of superstrings, instead of two. They have rather uninspiring names. There are the open strings, called type I, two types of closed strings called IIA and IIB, and two types of 'heterotic' strings that are somewhere in between, called HE and HO. The latter won't detain us, although they are indeed interesting. All these live in ten (nine space and one time) space-time dimensions.

Now, return to consider how we might link these ten-dimensional theories to our known four-dimensional space-time. We might do this by rolling up the extra six space dimensions into some sort of tiny space. This we can do but it creates a curved space, and as we have seen, string theory imposes some conditions in this case. One way to solve these conditions is by restricting the six rolled up dimensions to be a particular type of space called a Calabi-Yau surface. A two-dimensional slice of one of these is shown in Figure 4.2.

These spaces are very interesting geometrically. But a surprise awaited the researchers investigating this. They found that when you studied the IIA superstring theory rolled up on a particular Calabi-Yau space, and the IIB superstring theory on a certain, different Calabi-Yau space, the results were identical, even though the Calabi-Yau spaces were completely different. This symmetry was called 'mirror symmetry' and it spawned an entirely new area of research.

Mirror symmetry is rather similar to the symmetry we described earlier where string theory on a tiny circle is the same as on a large circle

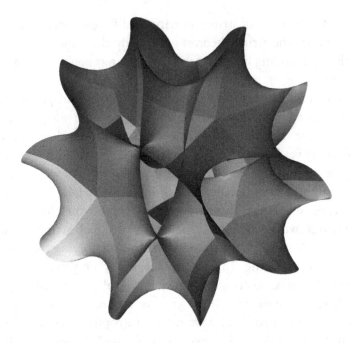

FIGURE 4.2   A slice of a Calabi-Yau manifold

with the inverse radius. In fact, when you put supersymmetry into this, the relationship links the IIA theory to the IIB one. This symmetry is called 'T-duality'.

Are there other ways in which two apparently different string theories are in fact the same? Indeed so, and a key idea here goes a long way back in time. The phenomena of electricity and magnetism have been observed for centuries, but the development of a precise theory to describe them was not found until around 150 years ago, when Maxwell formulated his eponymous equations. These unified electricity and magnetism into one formalism, which also opened up new advances. For example, wave solutions of these equations were found, and it was concluded that these describe the propagation of light. This was something which no one had previously imagined might relate to electricity and magnetism at all.

Something else was apparent in these new equations. If there was no matter, so you looked purely at the equations in a vacuum, then you could swap the electric and magnetic fields and the equations didn't change. Furthermore, if you allowed particles with magnetic as well as electric charge, then this 'duality' also held, provided that these charges were inversely related in a certain way. Such magnetically charged particles, or 'magnetic monopoles', have not so far been discovered, but they are not forbidden by the equations. Experience in physics has been that what is not forbidden is compulsory, so we keep looking.

A similar duality appears in string theory, called 'S-duality', and it implies that a certain string theory with a given coupling constant is equivalent to a different string theory with the inverse coupling constant. This is extremely interesting as it potentially addresses a key problem in quantum theory. We mentioned earlier that quantum particles scatter from one another when they meet, and that this is calculated using quantum field theory. In more detail, this calculation proceeds by an approximation method, which loosely speaking sees how many times there is interaction with virtual particles appearing out of the vacuum, and works out the results for each case. This gets much more complicated the more interactions take place, and to get the final full answer you have to keep summing up these additional pieces – which are infinite in number.

But you can only sum up an infinite set of numbers to get something finite if they keep getting smaller fast enough. For example, $1 + 1 + 1 + \ldots$ is going to add to infinity, whereas $1 + 1/2 + 1/4 + 1/8 + 1/16 + \ldots$ can be shown to add up to 2. You can quickly persuade yourself of this by adding more terms with a calculator, or by looking at Figure 4.3, where the first dot is the first term, the second dot is the sum of the first and second terms, etc.

The number that is linked in quantum field theory to the interactions mentioned above is called the coupling constant. This has to be small enough so that when you add all the interaction terms you get a finite answer, like in Figure 4.3. The magic that happens if there is an S-duality is that if the coupling constant is large, you can just switch to the equivalent S-dual string theory, for which the coupling constant is the inverse of that number. The inverse of a number is one divided by that number, so the inverse of a large number is a small number (e.g., the inverse of 100 is one hundredth). Thus, in the S-dual theory in this case one has a small coupling constant and one can work out the quantum theory. This has been shown to work in a number of cases. The original S duality was found to relate the IIA and IIB string theories, just like the T duality. Here is yet

| 1.0 | 1.2 | 1.4 | 1.6 | 1.8 | 2.0 |

FIGURE 4.3 $1 + 1/2 + 1/4 + 1/8 + 1/16 + \ldots = 2$

another new idea from string theory that sheds light, using radical new ideas, on problems in our current theories.

This might all sound pretty complicated, as we seem to have five different string theories, with some of them related to others by different sorts of dualities. There is in fact a whole web of duality relationships amongst the string theories. This raises some obvious questions. One is whether the string theories are really different at all, and similarly, whether the dualities are different. Might all of this fit within a structure where there is a 'Master' string theory, which has just one big duality symmetry? This is precisely what did emerge upon further exploration, with a dramatically new approach as we will shortly see. Within this, the challenge of finding the string geometry that can explain dualities involving different spaces has morphed into a new language.

There is much conjectured in relation to this, but what is certain from what we know now is that as far as string theory is concerned, putting it in simple terms, your donut may as well not have a hole, and how big it is a matter of debate.

CHAPTER 5

Brane Waves, or We're Just Blowing Bubbles

M AYBE YOU THOUGHT YOU could rest up a bit now, as you are getting to know a bit about string theory. That might be true if string theory was just a theory of strings, as we have described it so far, but it turns out that it is much more than this. An exceptional period of creativity, starting around the mid-1990s, revealed that we need to think about string theory in a much broader context, with wholly new ideas and approaches coming into play.

The work on closed strings had revealed evidence for a new geometry for string theory, and connections with gravity. People had also studied open strings, and it had been realized that you could attach charges, like electric charge, to the ends of these strings. The theories that arose from open strings were then similar to our successful theories of the electromagnetic and nuclear forces. The ends of open strings can also join to form closed strings, which might describe gravity. Thus, in sum, there is the possibility that string theory could be a unified theory of all the known forces of nature. It is this hope that has driven much of the research in string theory as it is seen by many as our best candidate for a unified field theory, a 'Theory of Everything'.

In the 1990s, it was found that the ends of open strings can do something else apart from join up, and this discovery had remarkable ramifications for the subject. To solve most equations in physics you need to put in some information about the system being studied, called 'boundary conditions'. For example, to predict where a cannon ball is going to land, you need to know its initial speed and direction of travel as it emerges from the cannon. These two pieces of information are the boundary conditions.

For open strings, the boundary conditions are that you need to say how the ends of the strings behave. One set of such boundary conditions ('Neumann') says that the ends move at the speed of light. There is also another consistent set of boundary conditions ('Dirichlet'), which dictate that the ends of the string are fixed in place. Not much attention was paid to this case initially, however, as it didn't seem very sensible to fix the ends like this. How could they be stuck in space and what would they be stuck onto?

The subject developed quickly when theorists explored what would happen if the places where the ends of the open strings could be fixed could form whole surfaces, with these surfaces being new and dynamical objects in the theory. These were called 'D-branes', with the 'D' referring to the Dirichlet boundary conditions and the 'brane' referring to the surface, thought of as like a membrane.

Whilst string theory was developing, there had been some parallel research on higher-dimensional analogues. Instead of studying how vibrating strings might behave, one can investigate vibrating surfaces more generally, that is, membranes. These theories had proved difficult to solve in general; however, they had been shown to be solutions of supergravity. Supergravity is the supersymmetric form of regular gravity, and it comes out of superstring theory. A whole new structure emerged as further research was done in this area, with diverse 'brane' solutions of string theory being found. The string itself is thought of simply as a one-dimensional brane in this framework.

The geometry of these branes is fascinating, and they can be thought of as analogous to higher-dimensional soap bubbles and films. The physics of soap films is rather exceptional. How can you make a liquid like water exist as a two-dimensional surface?! Their geometry is just as interesting. Of course, floating freely, soap films form into spherical bubbles – there is tension in the soap film that shrinks it, but the air inside the bubble is resistant to being compressed, and these two forces balance out when the bubble is spherical and of a suitable size. If you blow bubbles, then you sometimes get more complicated objects, such as two bubbles joined together. More generally, if you look at foam that is made from soap films, you will see what appears to be a complicated structure of cells. It turns out that soap doesn't just come as spherical bubbles, it can be squares and cubes, for example!

Although this appears complicated, there is an amazingly simple structure underlying the way soap films meet. There are only two rules:

Rule 1: soap films only meet three at a time along arcs at equal angles of 120 degrees and

Rule 2: these arcs only meet four at a time at a point, at a fixed angle (109.47.. degrees since you asked)

Figure 5.1 illustrates this. You should imagine wire frames to which the outside edges of the soap films are attached, a pyramid for the left-hand picture, and a cube for the right-hand one. Then, for example, the soap films on the inside of the pyramid all meet three at a time along arcs (lines) according to Rule 1, and the four lines all meet at one point in the middle according to Rule 2.

It doesn't matter what is happening on the outside. For example, for the cube picture on the right-hand side, if you look closely at this you will see that the two rules still apply to the soap films inside the cubical frame.

You can make these shapes yourself with wire frames and a soap and water solution, with a little gelatin added. You can create more complex shapes as well. For example, for the wire cube, if you are careful when you

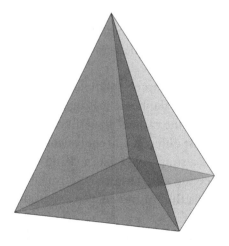
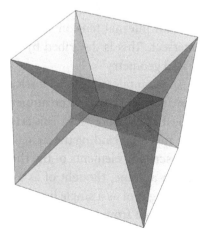

FIGURE 5.1 Soap films

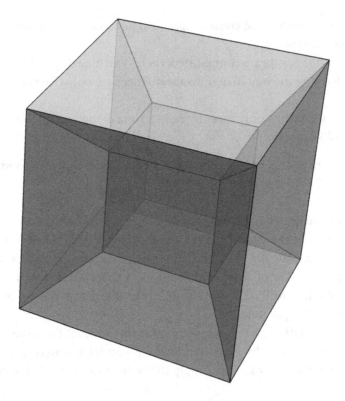

FIGURE 5.2   A cube made from soap films

pull this out of the solution you will sometimes get a cube inside the cube, as shown in Figure 5.2.

The 'branes' of string theory have analogous properties, as they also have an internal tension that leads to a set of rules about how they can intersect. This is described by rather beautiful mathematics called 'calibrated geometry'.

The complications and intricacies of all the brane solutions and how they interact proved to be numerous, but the take-home message appears to be 'brane democracy'. This is the idea that the various brane-like objects that appear, including the strings, are all just different and equivalent ways of describing elements of the theory. For a simple example, a string ending on a brane, thought of as a line meeting a surface, can equivalently be thought of as a single brane which has a thin funnel-shaped extension coming out from it (see Figure 5.3).

These ideas led to a classification of possible string theories, and the relationships between them. The five superstring theories mentioned in

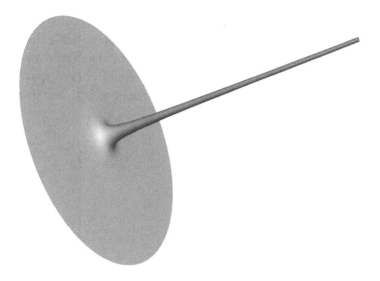

FIGURE 5.3   A string ending on a brane (both the flat surface and the funnel should be thought of as extending to infinity)

Chapter 4 are related by the duality transformations, which are closely related to the string topology and geometry we described earlier.

It wouldn't be very illuminating to detail the differences and relation-ships between these five theories here, except to note again that they all require ten-dimensional space-time (nine space dimensions and one time). What is notable, however, is that it was soon realized that these five theories could all be derived from one master theory, which lives in not ten but 11 dimensions. This theory was called 'M theory', with the letter standing possibly for membrane, magic, or mystery.

M theory is thus a proposed unifying theory for the different string theories that have been discovered. String theories describe vibrat-ing string-like objects in ten dimensions. M theory contains a theory of vibrating two-dimensional membranes ('M2-branes') living in 11 space-time dimensions. When the energies are low, an 11-dimensional supergravity theory emerges, but M theory itself is much more myste-rious. A fundamental object in the theory, a 'three-form field strength' which is linked to the M2-brane, appears to be pointing us towards another new sort of geometry, an 'M geometry' if you like, which could be the right framework for formulating the theory in a more funda-mental and powerful way. This is under active investigation. A prom-ising approach uses the ideas of 'doubled' or 'exceptional' geometry, where the space-time is a kind of projection of a larger construct which places the form field strengths on an equal footing with the metric which describes the space-time. If you thought that you'd started to get your head around imagining 11-dimensional curved space-time, then this new geometry, still under construction, will certainly provide some new challenges!

CHAPTER 6

You Are a Screen Idol

M OVIES ARE PROJECTED ONTO a screen in a cinema, or shown on a phone, laptop or TV screen. These are two-dimensional representations of three-dimensional worlds. With 3-D glasses or via holography, you can have three-dimensional effects. These sorts of effects, where a higher-dimensional world can be represented in lower dimensions, are rather rare in physics. But as usual string theory is an exception, and this might provide further key clues to the real nature of strings.

String holography started with a remarkable discovery in 1997. We have talked about supersymmetry, uniting force and matter particles, and combining this with string theory gives superstring theory (what else?). Superstrings can exist in ten dimensions if the space-times satisfy appropriate equations. A rather nice curved space-time of this type has half of the dimensions forming a five-dimensional sphere, and the other half forming something called 'anti-deSitter space', which maybe is best just thought of as a kind of stretched five-dimensional surface.

Spaces don't have to go on forever or be rolled up, as they can have boundaries. For example, the surface of a disk has a boundary which is the circle surrounding the disk on the outside, its perimeter. The anti-deSitter space just mentioned has a boundary, and this is a four-dimensional space. Keep this in mind for the moment, as the other part of this holography story involves a rather beautiful quantum theory which will now be described.

The theories describing the non-gravity forces are examples of 'Yang-Mills' theories. These are a whole class of theories of particles where the particles are formed into particular family groups, so that they are relatives of one another. The relationships between particles are called symmetries as we saw earlier, or supersymmetries if they relate force and matter particles. There are only a few types of supersymmetric Yang-Mills theories, labelled by an integer N, which can be 1, 2, or 4. Not much choice there, but since we are looking for a single unified field theory, the less choice the better generally. The best theory, for lots of reasons, is the N = 4 one, and it has been studied a lot.

Physics has had a perfect basic mathematical model for some time. It is called the harmonic oscillator, and it has a very nice solution. This very simply describes motion which repeats, for example, the swinging of a pendulum. The quantum version of this underlies all our quantum field theories. For the quantum harmonic oscillator the repetition is the 'frequency' of the particle, related to its wavelength when it is looked at as a wave.

The N = 4 supersymmetric Yang-Mills theory (what a mouthful, but nobody has given it a better name) might be viewed as a souped-up version of the harmonic oscillator, suitable for the 21st century. This is because it is rather like a perfect model quantum field theory, with all sorts of beautiful properties. It has lots of symmetries, as just mentioned, but it also has lots of rather secret, hidden symmetries that only emerge when you investigate it in certain ways, for example, in a kind of dual or mirror picture. The way particles in this theory interact is also much neater than how they interact in other theories. It is not a realistic theory in the sense of matching what is observed in experiments, but because it is much more amenable to analysis than other theories, we can learn a lot about new ways to investigate them using this model theory.

But now back to our story of superstrings in ten dimensions, looked at in the space which is a five-dimensional sphere combined with an 'anti-deSitter' space. What was found in 1997 was that the superstring theory on this ten-dimensional space-time could be matched to the four-dimensional N = 4 Yang-Mills theory on the boundary of the space. Since then other examples of this relationship have been found.

What this suggests is something profound, that gravity is a sort of holography machine and it projects itself onto lower dimensions as the other forces. A simple way to think about this is that a theory of gravity in some space-time can be equivalently thought of as a theory of force and matter particles on the boundary of that space-time. Figure 6.1 offers an indicative sketch. The interior of the half-sphere is where gravity operates, and the gravity is 'projected' onto the boundary surface, where it appears as the other forces and particles of nature.

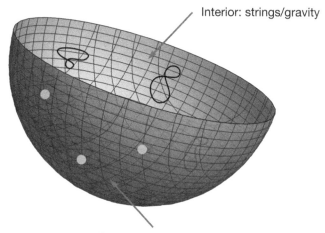

Interior: strings/gravity

Boundary surface: gauge theory/particles

FIGURE 6.1   The idea of holography

The obvious conjecture then is the extraordinary idea that perhaps our world of matter and forces is just a holographic projection onto our universe of a higher-dimensional gravitational world. We might all be just gravity holograms, projections of higher-dimensional gravity onto our world.

Let's Twistor Again

W E STARTED OFF BY talking about twistors, and how they were invented in the 1960s but failed to catch on in the physics community; a bit like the fashions of the period generally. But a few physicists kept working on this approach, perhaps whilst wearing flared trousers, and the theory was always there in the background, with the thought that this stuff really should come in useful one day. Then suddenly, in late 2003, it moved to centre stage.

The insight that led to this was a new understanding of a long-standing and striking puzzle. It was mentioned earlier that to make predictions for what goes on in key experiments, like those at the Large Hadron Collider for example, you need to complete extremely difficult calculations. If you directly apply the rules of quantum field theory to get the formulae to predict what will happen, you get very long and opaque expressions. For example, if you consider the lowest-level scattering of just five 'gluons', the particles that hold protons and other particles together, the result takes many pages to write out. Just one of these pages looks like Figure 7.1. We will spare the reader the situation with gravity calculations, as it is even worse! The symbols in this formula represent the speed and spin of the particles.

Figure 7.2 shows another formula, written in a different mathematical language. The many pages of symbols just mentioned can be shown to be precisely equal to this one-line formula. It is surprises like this that send out a signal loud and clear. If you are using the first approach, you are barking very much up the incorrect tree and you should be using the language of Figure 7.2. But what is this language? Here we just note for later reference that the angle brackets in that Figure refer to another way to describe the gluons, using 'spinors'.

FIGURE 7.1   A small part of a horrible formula for five-gluon scattering

$$A_5(1^-, 2^-, 3^+, 4^+, 5^+) = i\,\frac{\langle 12\rangle^3}{\langle 23\rangle\langle 34\rangle\langle 45\rangle\langle 51\rangle}$$

FIGURE 7.2   Much better

But whilst people could prove, by laborious calculation, that the complicated expression described above was mathematically equal to the simple one, it remained a mystery why such a simple formula should exist. And without really understanding what was going on underneath this, it was not possible to easily find other examples of such simple formulae.

The solution to this puzzle remained buried for over 15 years until in late 2003 when it was found that if you use the language of twistor space, then a simple geometric picture emerges which explains the simplicity above. The clue was that the spinor language of Figure 7.2 can be easily translated into twistors, which it turns out look rather like pairs of spinors. Each particle in the scattering process is described by a point in twistor space, and the equation describing the scattering is simply telling us that the points representing the particles all lie on a straight line. Thus, horribly complex expressions as illustrated by the excerpt in Figure 7.1 can be described, using the twistor language, by simple geometrical pictures involving points and straight lines.

This work led to a rapid expansion of new insights and techniques which continues to this day. One of the early steps was the discovery that you could use formulae like the one in Figure 7.2 as 'building blocks'. These blocks are called 'MHV amplitudes' and if you glue them together you get new, simple representations of quantum results. The simplest 'one-loop' diagram is shown in Figure 7.3.

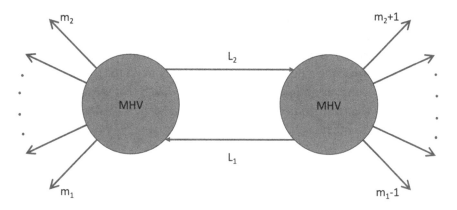

FIGURE 7.3   A quantum MHV diagram

Another early result was a discovery that probably should have been made decades earlier, as it used well-known facts. This was that you could use the general properties of how particles interact to predict the results for a process involving a certain number of particles interacting, from the results for fewer particles. Thus, you could get the scattering predictions for five gluons from the results for four and three gluons, for example. In the simplest case, for 'k' particles you simply attach a three-particle diagram (the blob on the left in Figure 7.4) to an (k − 1)-particle diagram

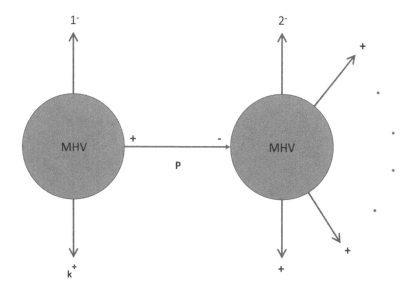

FIGURE 7.4   A recursion picture

(the right-hand blob in Figure 7.4) to get the scattering predictions for how 'k' particles interact.

These so-called 'recursion relations' have proved very powerful. First, in terms of providing a new and much easier way to do the tough calculations, and second, as an insight into a new approach that we will describe later, where the recursion corresponds to building new diagrams by adding 'bridges' to others.

CHAPTER 8

The Square Root of Gravity

O K, SO MUCH NEW and strange. And we haven't even got to the last ten years yet. Let's start on that. It's helpful for the subject of this chapter to think about 'spin'. Any everyday object can spin around, such as a top, a ball, or the earth. What about quantum particles? These can't be thought of as little spheres because of the wave/particle nature of quantum theory. This is the Heisenberg Uncertainty Principle which states, roughly speaking, that particles are 'fuzzy' objects and you can't in general know exactly where they are or where they are going. The principle relates the uncertainty in position with that in momentum. Nonetheless, you can make certain predictions about the probabilities of finding them with definite positions or momenta.

There is, however, another less well-known Uncertainty Principle, whose existence shouldn't be a surprise once you hear about it. Instead of uncertainty in the linear position and momentum of a particle, it relates the uncertainty in the angular or rotational position with the uncertainty in the speed of rotation. Whilst this is a rotational version of the linear Heisenberg principle, it is quite different in its consequences from the simple fact that if you rotate anything right around (i.e., by 360 degrees), then it ought to look the same as it did initially, even for a quantum particle. This led to the conclusion that all quantum particles have something called 'intrinsic' spin, which is only allowed to take certain values, that is, it is 'quantized'. The allowed values are 0, ½, 1, 1½, 2, and so on, in appropriate units. I like to think of this as the fuzziness of rotation. You don't know if a particle you are looking at is in some place, or if it has done one or more quick 360-degree rotations on the spot whilst you were blinking so to speak.

We mentioned that the graviton, which carries the force of gravity, has spin two. The particles that carry the other forces, such as the photon which carries the electromagnetic force, have spin one. This is another mystery, especially if you are trying to unify all the forces into one theory. One plus one is two; so maybe two photons glued together make a graviton? Sometimes the initial motivation for a whole new theory of physics does indeed come from very simple-minded ideas like this. People tried this idea out in various ways over the years without much success. But, in another of the extraordinary inspirations coming from string theory, in the past ten years an entirely new way to investigate this has come about with more than just mathematical applications.

The interaction of quantum particles is called 'scattering' which describes particles approaching each other, colliding or interacting, and then shooting off again, possibly as a different set of particles. The calculations of quantum field theory tell us how likely any particular scattering process is. In the Large Hadron Collider for example, protons are smashed into each other head-on at high speeds, and annihilate and explode into lots of other particles that are then detected. The challenge is to see if there are new, more fundamental particles amongst all of these, or other new particles that will provide clues for how to improve our fundamental theories of matter. This is how the Higgs boson was discovered. As noted earlier, the calculations needed in quantum field theory rapidly become incredibly difficult to do when there are more than a few particles involved, and this is a major problem. But we have already seen evidence that this is another problem that string theory promises to solve, and there is more to come.

When you work out the formulae for the different sorts of scattering of spin one particles, these are all different of course, but they all have the same general structure. They are sums of certain coefficients depending on how fast and in what direction the particles are going, and how their spin is oriented, multiplied by numbers that relate to the charges of the particles.

The formulae for the scattering of the spin two gravitons are different, and even more complicated, due to the fact that, in essence, the higher spin allows for more things to happen. The answers look like sums of complicated expressions depending on the motion of the gravitons. These formulae really don't look much like those for the spin one particles.

However, in 2008 a rather magical trick was found that changed this situation entirely. A way was found to rewrite the spin one scattering formulae,

$$A^{spin1} = \sum \frac{N \times C}{\prod_i D_i} \qquad\qquad M^{spin2} = \sum \frac{N \times N}{\prod_i D_i}$$

FIGURE 8.1   A spin one formula   FIGURE 8.2   A spin two formula

let's call them 'A', in such a way that they still took the same general form of a sum of terms as before. Let's call these N times C, where the N piece contains the speed and directions of the particles and the C piece contains the charge-related terms. These formulae look schematically like the one in Figure 8.1.

(For the more curious, the sum is over certain Feynman diagrams and the product is over the particle 'propagators' describing how they move. The latter are the D terms in the formula shown in Figure 8.1).

This sum of terms, although equivalent to the usual way of writing the formula, looked different. It had the extraordinary feature that if you just replace all the C's by N's, getting a sum of terms of the form given in Figure 8.2, then you got precisely the formulae for the scattering of gravitons. (Hence, the 'spin two' in the formula). The simplest example is the fact that if you take the formula for how three gluons interact, on the left-hand side of Figure 8.3, then the expression for how three gravitons interact (on the right-hand side) is just the square of this.

$$A_3 = \frac{\langle 12 \rangle^3}{\langle 23 \rangle \langle 31 \rangle}, \qquad M_3 = (A_3)^2$$

FIGURE 8.3   Three gluons/gravitons

In simple terms, if the scattering equations for the spin one particles, describing the non-gravity forces, contain terms called 'N', then the equations for gravity contain terms N times N, that is, N squared. The other forces of nature might best be thought of as the square root of gravity!

This simple observation has had a wide range of consequences. One is that this 'squaring' procedure can be generalized and leads to a 'web' of different expressions describing the scattering of other sorts of particles. An obvious question to ask is, for example, what happens if instead of replacing the C's by N's in Figure 8.1, you replace the N's by C's? The answer is that you get formulae describing how spin zero particles interact. Thus, the formulae for describing how various sorts of particles interact are more unified into one approach.

Another impact of this discovery is that it greatly facilitates calculations. It was noted in Chapter 7 that the spinor language of Figure 7.2

reduced a huge complicated expression to something that you could write on one line. But there were still challenges remaining to find the spinor formulae in many cases. One such challenge is deriving expressions for the more complicated graviton interactions, and the new 'square/square root' approach (now called the 'double copy') has greatly simplified this. More recently, this double copy has been shown to provide simplified descriptions of solutions to Einstein's equations for gravity. For example, the space-time outside a black hole looks like a double copy of the electric field around a point charge.

This is not just of theoretical interest, as one of the major achievements of our age in astronomy is the detection of gravitational waves. This opens an entirely new window to the universe. We have previously relied purely on studying the light waves from distant objects in order to explore the structure of the universe. It is extremely difficult to detect gravitational waves, as their effects at great distances are very small, but for cataclysmic events like the merger of two black holes, so much gravitational energy is emitted that the waves reaching us can now be detected by very sophisticated equipment.

We can use this information to test Einstein's theory of gravity, and the possible quantum corrections to it in order to understand in more detail how these massive events take place. But this requires us to perform gravitational wave calculations. Very recent work has used results from the double-copy approach in order to obtain new predictions for these processes.

Finally, where does string theory come into all this? Well, developments in string theory motivated the early work on the double copy. It was already known that the formulae for open string interactions had elements that looked like the square root of parts of the formulae for closed string interactions. These implied certain relations between the formulae for gluons and those for gravitons, inspiring the research described above. The full implications in string theory itself are still being worked out, and this is expected to lead to new insights and progress.

CHAPTER 9

It's Only Platonic

LIKE US, HUMANS IN the past of course wondered about the nature of the universe, and ancient civilizations had their own Theories of Everything. The ancient Greeks proposed a theory of the four fundamental elements earth, water, air, and fire, with all matter made from these (later Aristotle added a fifth element, the 'aether'). The properties of different sorts of matter were understood as reflecting the proportions of these elements in the matter.

These ideas were also reflected in their theory of human character, the 'humours' or fundamental fluids of medicine, and the seasons. Fire, linked to yellow bile, Summer and a choleric temperament; air, linked to blood, Spring and a sanguine approach to life; earth, linked to Autumn, black bile and a melancholy nature, and finally water, associated to Winter, phlegm (of course), and a phlegmatic character.

The ancient Greeks were also keen on symmetry, and in particular what we now call Platonic solids. Plato was a philosopher who argued that there was a link between the most symmetric and beautiful shapes and the fundamental elements of the Greek theory of matter.

The sort of shapes they were interested in were those you can make by gluing together simple polygons (triangles, squares, five-sided pentagons, six-sided hexagons etc.) to make closed surfaces in three dimensions. You can join four triangles and a square along their edges to make a pyramid for example, like those in Egypt, where the base of the pyramid is a square and the four sides are triangles. Or you can join six squares along their edges to make a cube. An example is a die and its six sides. The points where the polygons meet (e.g., the corners of a cube where three

FIGURE 9.1 The Platonic solids

squares meet) are called 'vertices'. The most symmetric shapes are those where you only use one type of polygon and all the vertices look the same. You can make only five of these special shapes. These are the cube, tetrahedron, octahedron, dodecahedron, and icosahedron, and these are called the Platonic solids (Figure 9.1). They are made from four, six, eight, twelve, and twenty polyhedral faces, respectively (e.g., the icosahedron is made from 20 triangles glued together).

Plato thought that these beautiful shapes ought to underlie nature, and he associated the fundamental elements with these solids. Fire was linked with the tetrahedron, earth with the cube, air with the octahedron, and water with the icosahedron. The dodecahedron was linked with arranging the constellations in the heavens. The solids and the elements they were linked to had common properties. The sharp points of the tetrahedron correspond with the sharp pain from fire's heat, and the more ball-like icosahedron has flow-like properties like water, for example.

The symmetry and beauty of these shapes also made them attractive as possible elements in much later theories of nature. For example, the astronomer Kepler formulated a theory of the solar system in the 16th century, based on relating the positions of the five known planets at that time to a model based on the five Platonic solids. These investigations, although unsuccessful, later helped lead to his landmark three laws of planetary motion, which had a profound influence on the subsequent development of astronomy.

Quantum field theory, with its vastly complex array of formalism and computational challenges, is perhaps the last place you would expect to find any relationship with the almost kindergarten-like process of gluing together polygons into pretty symmetric shapes that we have described above. However, this is what happens, and it provides another extraordinary clue as to what language we might best use to understand our current physical theories and where we may develop them next.

This is a recent story, emerging within the past ten years or so. Scattering amplitudes were mentioned earlier, which are the fundamental calculations in quantum field theory that tell you how quantum particles interact. There have been dramatic developments in the study of scattering amplitudes recently, which will be described in Chapter 10. These have led to new formulae, equivalent to but often dramatically simpler than previous answers. These new formulae arise from different approaches to the calculations, and raise questions about new interpretations of the answers. For example, in many cases one can derive two apparently completely different formulae for the same scattering amplitude, which you can miraculously prove to be equivalent, but only by laborious calculation. As we saw in Chapter 7, these situations in physics always suggest that there is a better picture, a new formalism or approach, within which these mysterious equivalences become clear consequences of this new paradigm.

An example of this, which was an early inspiration to subsequent progress, concerned formulae for the scattering of six gluons. Gluons are the particles that bind together the quarks inside the proton and neutron. Two different formulae could be derived, but their equivalence was not at all clear as they looked completely different. There are important features of amplitudes called 'poles', which are values of the variables near which the amplitudes become large. But these poles, which must be present, were not at all obvious in these formulae, and what is more there also appeared to be incorrect poles present.

The resolution of these apparent problems came very quickly when the right picture was found. This turned out to involve higher-dimensional polyhedra called polytopes. The special polyhedra, called Platonic solids that we just met, can be made from gluing together polygons along their edges to make a closed surface. The 'faces' are these two-dimensional polygons (e.g., the faces of the cube are squares). Similarly, one can glue together three-dimensional polyhedra to form a higher-dimensional closed surface called a 4-polytope, which is much harder to picture but is mathematically well-defined. The 'faces' of the 4-polytope are the three-dimensional polyhedra, analogous to the faces of a 3-polytope (like the Platonic solids) being two-dimensional polygons.

The remarkable observation that clarified the question above about different formulae for the six-gluon scattering amplitude was that this amplitude corresponded to the volume of a particular 4-polytope. The different ways of expressing the amplitude were simply different ways of calculating the volume of this higher-dimensional geometric object by gluing together simpler pieces whose volumes were known directly. This is quite extraordinary, as how could it be that the probability amplitudes for quantum particle collisions could be described by volumes of what you might think of as higher-dimensional cousins of the Platonic solids of the ancient Greeks?

It does happen sometimes that surprising results in physics are just coincidences, and aren't clues that there is something more profound going on. What was needed was to see if the above intriguing result about amplitudes could be generalized. This was done, and it was indeed found that other scattering amplitudes could be written in such a way that they were just given by the volume of higher-dimensional geometric solids.

These first results concerned the simplest scattering amplitudes, called 'tree' amplitudes, for which the quantum nature of the particles is suppressed. If you allow quantum effects, you obtain 'loop' amplitudes, with the word loop referring to the creation and annihilation of 'virtual' quantum particles. One might think of this in simple terms as being allowed by the Uncertainty Principle. The loop amplitudes are where the infinities of quantum field theory come in, discussed earlier in Chapter 2.

Quantum, or loop scattering amplitudes are much more difficult to deal with than their tree-like cousins, so it was not at all clear that the geometric picture of tree amplitudes that we have just described would work for their quantum versions. However, explorations of this did indeed find evidence that certain classes of quantum scattering amplitudes, describing the collisions of quantum particles, could be described by looking at a beautiful geometric object and simply calculating how big it is, or in other words, its volume. It is almost as if the Greek theory of matter has returned in a sophisticated guise. This higher-dimensional geometric object is a still rather mysterious object which has been named the amplituhedron. Work is underway exploring more about this.

One thing I left until last ... the amplituhedron lives in momentum twistor space, which brings us back, in Chapter 10, to the story that we started this book with ....

CHAPTER 10

Everything Is
Now – Now

I N CHAPTER 1, I explained how twistor space provides a framework for physics which is profoundly different from space-time, replacing the precise points of space-time (events at a given place and time) with light rays. In the original formulation it proved difficult to re-write our successful quantum field theories in this new language, or similarly to describe what quantum gravity might be. As a consequence, the exploration of the possible use of twistor space in physics remained a minority interest.

String theory, as described in subsequent chapters, has revealed entirely new viewpoints. Amongst these, one of the most unexpected has been the discovery that the formulae that describe the interactions of quantum field theory, scattering amplitudes, are dramatically simplified if one does not use the language of space-time, but instead introduces the twistor formalism. A recent and extraordinary result is that each scattering amplitude is simply giving the volume of a higher-dimensional geometric space akin to a Platonic solid, the amplituhedron. Physicists are very enthusiastic about anything that simplifies their formulae. This is immediately of practical benefit as it is much easier to work with simpler equations, but more fundamentally, simpler formulae usually reflect a more profound and powerful framework for understanding and developing the theories that describe nature.

The formulae for these volumes associated to scattering amplitudes are written using a form of twistor space. This reveals that Penrose was right after all to think that his new space of light rays could illuminate our understanding of quantum physics. There is a lot that can be said about this new twist in the tale, but let me focus on one key aspect.

The expressions for the amplitudes have an equivalent and beautiful geometric formulation in terms of spaces of higher-dimensional planes or sheets, called Grassmannians. These are a bit like the soap films or brane geometries described earlier. The Grassmannians have a nested 'stratified' structure, where interesting things happen when you go towards the boundaries of subspaces and surprisingly, all of this corresponds directly to physical properties of particle scattering.

Another, and in many ways more profound, aspect of this formulation is that it appears to junk two of the most fundamental requirements of quantum physics, which are necessary for it to make sense. One of these is called 'unitarity', which can be understood in simple terms as saying that in every interaction of particles, something definitely has to happen, including the possibility that nothing happens. (Who said theoretical physics was hard to understand?) Quantum interactions can have different and unpredictable outcomes. What is predictable, however, is the list of possible outcomes and the probability that each of these will occur. What absolutely must happen is that the sum of those probabilities adds up to one. It has to be true that one of the possible outcomes must happen. And indeed, in every actual experimental test, this is found to be true. A number of experiments in the past seemed to find results that violated this, but in every single such case it was subsequently found that some of the possible outcomes had been missed, sometimes through some problem with the experimental set-up, and sometimes because the thing missed was not known before, for example, a new particle. In fact, looking for this 'missing probability' is an important way to look for new phenomena.

The other fundamental tenet of quantum physics that the Grassmannian description of particle interactions appeared to ignore is called 'locality'. This requires that the interactions of quantum particles occur at definite points in space at definite times; this is closely linked with 'causality', whereby events in the future are not allowed to influence events in the past.

The Grassmannian approach does, in the end, respect these fundamental requirements, as it gives the correct, known answers for quantum processes, although doing this in a completely different way. Unitarity and locality are much more self-evident in the traditional approach, and are built into it, whereas in the Grassmannian formulation, whilst they are still guaranteed, they are hidden and more appropriately thought of as emergent properties.

But if these basic ideas are not the ones evident in this new framework, what are the essential and evident features? Well, there are symmetries of the Grassmannians; you can rotate an infinite flat sheet about and it still looks the same for example, but these are not the immediate answer to this question. A key understanding comes when you seek to actually calculate in this approach. The original expressions are very simple to write down, but much more difficult to work out in general.

However, there are well-defined ways to proceed, and the steps involved can be translated into pictures that bring our story full-circle. These pictures are called 'on-shell diagrams' and they tell us that you can understand how real massless quantum particles interact by using diagrams that only involve other real massless quantum particles. What is more, the diagrams that you use are networks of points linked together, rather like the neurons of the brain linked by axons. These on-shell diagrams, remarkably, are made entirely from only two types of trivalent vertices (points with three lines emerging), conventionally coloured black or white. These three-point vertices are themselves well-known quantum amplitudes for the scattering of three particles; the lines are the particles and the vertex is the interaction.

A simple example is when two gluons collide and then emerge with different speeds and directions of travel. The simplest diagram describing this process is shown in Figure 10.1.

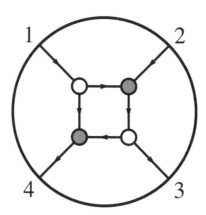

FIGURE 10.1   An on-shell diagram for four particles at tree level

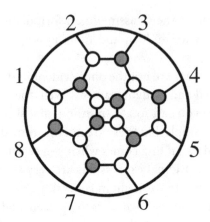

FIGURE 10.2   An on-shell diagram with eight particles

There are more complicated diagrams when you have more particles or if you take into account higher-level quantum effects. An eight-particle example is shown in Figure 10.2.

The 'recursion relations' mentioned in Chapter 7 have a very simple representation in this approach – they are just adding linked pairs of vertices ('bridges') to these diagrams in a well-defined way. This provides a simple way to generate and calculate results for the more complex diagrams from those for the simplest ones.

The whole approach here is different from that adopted by standard quantum field theory, where the Feynman diagrams have 'loops' of virtual particles. With this new formalism, all the particles involved are 'on-shell', meaning that they are real particles. They are also massless, so that everything in this approach uses particles travelling at the speed of light whose time, mathematically, appears to stop. In summary, for the particles in this theory, Everything Is Now. Which, appropriately, is where we started a while ago.

CHAPTER 11

Accentuate the Positive

THE PROBLEM WITH STRING theory, even when writing a telegraphic survey like this, is that events rapidly overtake you. There are many smart people working on the theory, and new interactions taking place with both experimental and mathematical physicists. As a result, a whole new area can open up and quickly grow. Let me finish by outlining some aspects of a very new development that is showing great promise. This is how the theory of positive geometry is providing wonderful new insights into our theories.

Everyday ('natural') numbers can be positive, negative, or zero. The positive numbers have some obvious features, such as if you add or multiply two of them the result is also positive, and there is a 'boundary' to the space of positive numbers, the number zero. You can get arbitrarily close to zero with smaller and smaller positive numbers. These very simple facts have analogues in many areas of mathematics.

It turns out that geometry can be positive as well as arithmetic. For example, the interior of a triangle can be described by giving a set of numbers which are required to be positive, and the edges of the triangle, its boundary, correspond for each edge to one of these numbers going to zero. Then, you can glue triangles together to form polygons, gluing two to form a square, three for a pentagon, and so on.

This is the sort of stuff that mathematicians like to do, and then they can ask questions like 'How many ways can you glue triangles together to form a polygon with "n" sides'? This number is called the Catalan number and equals two for the square, 5 for the pentagon, 14 for the hexagon, and so on, as per Figure 11.1.

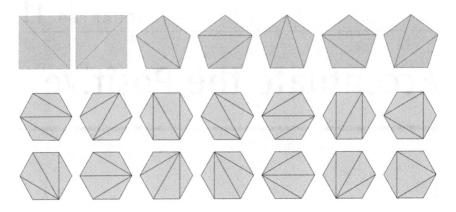

FIGURE 11.1   Different ways to glue triangles

How can this be relevant to what we have seen so far of string theory, and our understanding of physics? Well, first, the Grassmannians mentioned in Chapter 10 can be described using arrays of objects called matrices, and matrices have numbers associated to them, called 'minors'. It turns out that the sets of matrices with particular sets of minors which are positive numbers are crucial to the study of the amplituhedron and its linkage with physical properties of particle scattering. In addition, the boundaries of these sets, which are subspaces of the Grassmannian, play an important role relating to physically interesting limits of these scattering processes. But going into this in more detail is going to get (even more) abstract and technical. Luckily, thanks to some recent work, we have a get-out clause as we will now see.

The most obvious place where being positive is essential in physics is with energy, as the energy of physical systems is always positive. In particular, this is true of the individual particles described earlier. This has particular consequences in the scattering of particles, as it imposes conditions on the incoming and outgoing particles. So far, so old hat. But research in 2017 discovered a new and surprising geometry, present even in very simple scattering processes involving only one sort of primitive particle.

Thus, let's think about the simplest interactions of the simplest possible particles, the ones without any spin, called 'scalars'. Consider how four of these particles might interact. This can be described by two particles approaching each other, an interaction taking place, and two particles later emerging. When two of these particles collide and interact, the process can be described by giving a number related to the energies involved,

FIGURE 11.2   An interval from 0 to C

which is bigger than zero and less than some positive number 'C'. Hence this number lies on the interval depicted in Figure 11.2.

This interval and its boundary points at zero (0) and 'C', have a natural correspondence with how the particles interact, as given by the so-called Feynman diagrams below the line in Figure 11.2. The lines in these diagrams represent the positions of the particles as they travel, and the particles can interact by merging into a new single particle which then later splits up into two again. The end-points represent two ways this merging/splitting can happen, and these are related by the 'X' shaped diagram where four particles interact at a point, which corresponds to the interval between the points zero and C.

This picture generalizes to describe how more than four particles interact. For five particles the picture is a polygon with five sides, and for six particles it is a surface you can draw in three dimensions, made from gluing three rectangles and six pentagons together. (See Figure 11.3. The colours or shading in this figure are purely decorative.) For more than six particles the surface lives in dimensions higher than three.

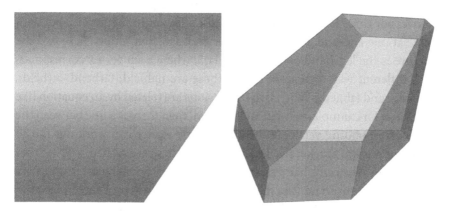

FIGURE 11.3   The five and six particle pictures

All of these surfaces have the properties that the energy variables of the scattering processes that they represent are positive in the regions inside any face of the surface, and all the boundaries of the faces correspond to one or more of these energies going to zero, with a corresponding special sort of particle interaction diagram.

It was very surprising to find such a geometric structure arising in the description of the simplest interactions of the simplest particles that we know of. What was even more surprising was that all of this structure had already been explored by mathematicians, for completely different reasons. They had called these shapes associahedra, simply because the pictures just correspond to all the different ways to put brackets around some list of letters. For example, for three letters 'a', 'b', and 'c' you can have '(ab)c' and 'a(bc)'. You might think of these expressions as representing the multiplication of numbers – in the former case you first multiply a times b to form the product ab, then multiply this by c. In the latter case, you first multiply b times c to get the product bc, and then multiply this by a. For ordinary multiplication of numbers, these are always equal, that is,

$$(ab)c = a(bc) \qquad (1)$$

For example, you get 24 from either (2 times 3) times 4, or 2 times (3 times 4). This property is called 'associativity'.

The mathematical picture of the associahedron is obtained by making a vertex for each different way to put a set of brackets on the letters, with a line, or edge, connecting any two vertices for which the two corresponding bracketed expressions are related by the simple property in Equation (1) above. For three letters this picture is just a line as in Figure 11.2 but with (ab)c at one end and a(bc) at the other end. For four letters, a, b, c, d the picture is the left-hand one in Figure 11.3, where each vertex corresponds to a different way to bracket these. These are ((ab)c)d, (a(bc))d, a((bc)d), a(b(cd)), and (ab)(cd). When two expressions are related by an equation like (1) above, for example, ((ab)c)d and (a(bc))d, then there is a line between the two corresponding vertices. The enthusiastic reader can work out the 14 different ways to bracket five letters and how these are related, and see that this matches the right-hand picture in Figure 11.3. Spoiler alert: the answer is below in Figure 11.4 (the matching is such that each of these objects is made from three rectangular and six five-sided faces glued together in the same way. The sizes and shapes are not relevant here).

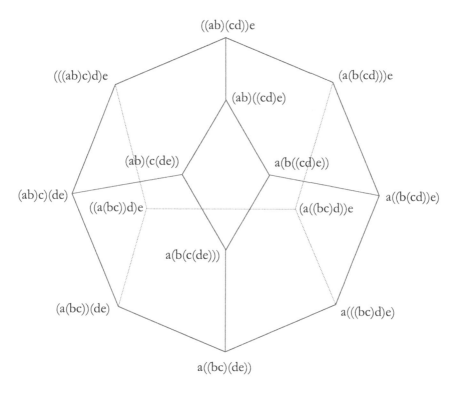

FIGURE 11.4   The associahedron for five letters

At the time of writing, this new geometric description of particle scattering is being rapidly applied to new areas and is bringing in a variety of
new and old mathematical techniques and approaches in new ways. It is
much too early to say where these paths will lead, but it seems clear that
there is much potential for this to be a new unifying approach to diverse
properties of quantum field theories and string theory.

CHAPTER 12

The Future

E VEN IF EVERYTHING IS NOW – now, there is still something to say about tomorrow. Most of the radical ideas introduced in earlier chapters continue to be explored and their applications developed. We still don't know what 'string theory' really is, in the sense of having a description that ties together all the astonishing features that have been discovered. Although we know a great deal about their properties, what the strings, or more generally branes, are 'made of' is a basic question that is not really answered yet. This seems to require some new sort of geometry, quite unlike any which we currently understand intuitively. This will need to explain in a unified way the properties we have seen, such as how big can be the same as small, how holes in space-time may be chimerical, how secret dimensions can hide next to us, or lie tightly wrapped up at every point, how strings might morph into membranes and back, how our world might be a gravitational hologram, how our everyday forces might be square roots of gravity, and what role positive geometry might play. And finally, how this all might fit into a description where everything is massless, time has stood still, and strange higher-dimensional geometric polyhedra stand as representations of the quantum world.

CHAPTER 13

String Theory and Reality

To FINISH, WE'VE SEEN plenty of wild ideas in this book, but what about reality? A perfectly fair question. To answer this, let's be clear at the start – a theory is just a set of conjectures until it makes predictions which can be repeatedly and reliably confirmed by experimental tests. And is there any experimental evidence for string theory? No, there isn't.

Are we then all just playing around with mathematics? Here the answer is both yes and no. Yes, obviously. But also, no, and there are two main reasons for this qualified answer. One is that much of what has been described in the earlier chapters does link up with current theories of physics, the challenges they face, and emerging experimental results. A recent example that we mentioned in Chapter 8 was the use of the 'square root of gravity' formalism to make predictions for corrections to the sort of gravitational wave signals that experimentalists are starting to observe. There are quantities like this that are or soon will be needed by experiments but are currently very difficult to calculate with known formalisms and approaches. More generally, string theory-inspired ideas have led to new and tremendously simplified expressions for scattering amplitudes, which are crucial to quantum field theory. You might take the view that these are just calculational aids, but it's pretty weird that these radical ideas enable us to match known calculations and then to go far beyond this. This strongly suggests that these new approaches are on the right track.

The other reason why many of us believe that it may turn out that we are not just doing mathematics is the one mentioned in the Introduction. The challenges that modern fundamental physics is facing are great, and rather simple-minded applications of current approaches show little prospect of addressing these. The problem of finding a consistent theory of quantum gravity is the most obvious example. This is not some arcane issue, as it is critical to understanding the black holes that live inside galaxies in the universe and the collision and merger of black holes that has recently been observed. It is also of course critical to understanding more about the origin and fate of the universe, what happened at times very close to the Big Bang, and how the universe will evolve in the distant future.

What we have seen in the past 30 years or so is the exceptional success of theoretical physics in providing a model of three of the four fundamental forces, the so-called 'Standard Model', which has been verified by all experimental tests done so far. These experiments are unparalleled in history in terms of complexity and technological challenges. Consider for just a minute what the famous Large Hadron Collider actually does. It accelerates 2,800 bunches of 115 billion protons in opposite directions to 99.999999% of the speed of light, guiding them around a 27-km tunnel 11,245 times per second before smashing them together with 600 million collisions per second, then processing 100 gigabytes of data per second to pick out the one in a million collisions of interest, and sending the data out to 140 centres around the world for analysis. Phew! It's hard enough reading this let alone imagining what it takes to actually do it. This sort of complexity and technological challenge is also a feature of other experiments relevant to high-energy physics, in modern astronomy.

Projects in fundamental experimental physics are becoming increasingly expensive and difficult, requiring larger collaborations and longer lead times. Recent experiments have confirmed our predictions, rather than finding unexpected new phenomena. If our current theories were faultless, this wouldn't be an issue and we could congratulate ourselves on our cleverness in devising a successful Theory of Everything. But our theories are far from complete, as noted above. They are also not pretty. The Standard Model is the sort of do-it-yourself job which is o.k. and functional, but you don't always like looking at it and you know you really will have to sort it out sometime. For example, it takes two full pages to properly define and write down the starting point of the Standard Model, before even doing anything with it, and it contains an abundance of constants that have to be determined by experiment rather than given by the theory. And even General Relativity, an astonishingly beautiful theory of classical gravity, fails utterly to transfer to the quantum world.

There are thus huge challenges remaining, and tremendous opportunities to advance our understanding of the universe by solving the great puzzles that still face us. The ideas inspired by string theory provide fresh, powerful new tools and extraordinary and radical new conceptual approaches. The brief summaries in this book provide just a glimpse of some of this new landscape. I hope that, as well as appealing to the simply curious, this book inspires some of the brightest minds of tomorrow to take up the challenge to answer the profound questions that we still face.

CHAPTER 14

References and Further Resources

I HAVE CLASSIFIED THE INTENDED audience level for the suggestions below by [P]: popular, with little technical material, [UG]: aimed at undergraduate physics level or above; may contain some popular-level material, [G]: graduate physics level or above.

CHAPTER 1: INTRODUCTION

Conlon's book 'Why string theory?' (CRC Press 2015) [P] is a longer review including some of the topics described here; see also the website http://whystringtheory.com. Of the somewhat older texts covering earlier material, Greene's 'The Elegant Universe: Superstrings, Hidden Dimensions and the Quest for the Ultimate Theory' (Vintage 2000) [P] was a best-seller and covers a breadth and depth of material in a readable style. Gubser's book 'The Little Book of String Theory' (Princeton University Press 2010) [P] provides a concise and readable survey including black holes, strings, branes, and dualities. Two others at a popular level are Jones and Robbins' 'String Theory for Dummies' (For Dummies 2009) and Musser's 'The Complete Idiot's Guide to String Theory' (Alpha 2008).

There are a number of texts at a physics or maths graduate student level. These include Johnson's 'D-Branes' (CUP 2006), Becker's 'String Theory and M Theory: A Modern Introduction' (CUP 2006), Dine's 'Supersymmetry and String Theory: Beyond the Standard Model' (CUP 2nd edition 2015), West's 'An Introduction to Strings and Branes' (CUP 2012), Ibanez and

Uranga's 'String Theory and Particle Physics: An Introduction to String Phenomenology' (CUP 2012), Zwiebach's 'A First Course in String Theory' (CUP 2009), Lust and Thiesen's 'Lectures on String Theory' (Springer 2014), and Polchinski's two volume 'String Theory' (CUP 2005). An historical summary of the origins of string theory is 'The Birth of String Theory' by Cappelli, Castellani, Colomo, and Di Vecchia (CUP 2012).

There are cosmological applications of string theory such as the multiverse and associated work, that are not described here. See, for example, Tegmark's 'Our Mathematical Universe' (Penguin 2014) [P]. Greene's 'The Hidden Reality: Parallel Universes and the Deep Laws of the Cosmos' (Penguin 2011) [P] focusses on the multiverse description of the universe and also covers M theory. His TED talk on string theory is at:

https://www.youtube.com/watch?v=YtdE662eY_M.

There is also work applying string theory to condensed matter physics – see Nastase's 'String Theory methods for condensed matter physics' (CUP 2017) [G] or the article:

https://www.quantamagazine.org/taming-superconductors-with-string-theory-20160121 [P].

There are other approaches to quantum gravity and applications. Rovelli's 'Reality Is Not What It Seems: The Journey to Quantum Gravity' (Penguin 2017) [P] has a summary of progress in loop space quantum gravity. See also his 'Seven Brief Lessons on Physics' (Penguin 2016) [P]. You might view the book you are reading as a string theory rejoinder to the latter.

Finally, some critics have argued that string theory has drifted too far from physics as an empirical science – for example, Hossenfelder's 'Lost in Math: How Beauty Leads Physics Astray' (Basic Books 2018) [P], Smolin's 'The Trouble with Physics: The Rise of String Theory, The Fall of a Science and What Comes Next' (Penguin 2008) [P], and Woit's 'Not Even Wrong: The Failure of String Theory and the Search for Unity in Physical Law' (Basic Books 2006) [P].

CHAPTER 2: EVERYTHING IS NOW – THEN

Some of the motivations for inventing twistors are described in: 'On the Origins of Twistor Theory', R. P. Penrose, from 'Gravitation and Geometry', a volume in honour of I. Robinson, Biblipolis, Naples 1987. See http://users.ox.ac.uk/~tweb/00001/index.shtml [UG]. A hand-drawn picture of the

Robinson congruence is in this reference. This is depicted in Figure 2.1 and on the front cover of this book. These figures and a number of others in this book were drawn using Mathematica®. A review of twistor theory is https://arxiv.org/pdf/1704.07464.pdf [G] and a mathematical text is 'An Introduction to Twistor Theory' S. A. Huggett and K. P. Tod (CUP 1994) [UG].

CHAPTER 3: IT'S RIGHT BEHIND YOU

'Extra Dimensions and Warped Geometries', L. Randall, Science 296 (2002) 1422 [P]. See also:

> http://citeseerx.ist.psu.edu/viewdoc/download;jsessionid=CE0172AC6E
> 484E016F566DBEFE58BFF9?doi=10.1.1.177.2161&rep=rep1&type=pdf

A review of large extra dimensions is:

> https://arxiv.org/pdf/0907.3074.pdf [UG].

CHAPTER 4: MY DONUT HAS NO HOLE

A short video on this area is at:

> https://plus.maths.org/content/geometry-strings [P].

A review is 'Stringy Geometry and Emergent Space', M. Marino: https://www.researchgate.net/publication/295902900_Stringy_geometry_and_emergent_space [UG].

Figure 4.2 is by Andrew J. Hanson -

> https://commons.wikimedia.org/w/index.php?curid=30579072

(OTRS Ticket#2014010910010981, CC BY-SA 3.0,)

CHAPTER 5: BRANE WAVES, OR WE'RE JUST BLOWING BUBBLES

'The Geometry of Soap Films and Soap Bubbles', F. J. Almgren and J. E. Taylor, Scientific American July 1976 [P].

Two interviews with Witten on dualities, M theory and related topics are at:

> https://www.quantamagazine.org/edward-witten-ponders-the-nature-of-reality-20171128/ [P], and

> https://www.ams.org/notices/201505/rnoti-p491.pdf

(the latter is more focused on mathematical aspects) [P]. See also Duff's article at:

> https://www.newscientist.com/article/mg21028152-400-theory-of-everything-have-we-now-got-one/ [P].

CHAPTER 6: YOU ARE A SCREEN IDOL

A recent popular talk, covering black holes and quantum entanglement is Brian Greene's Centre for Inquiry presentation 'The Nature of Space and Time' at: https://www.youtube.com/watch?v=M22MEShcyx8

'Understanding the Holographic Universe' on PBS space-time comprises 8 videos:

> https://www.youtube.com/playlist?list=PLsPUh22kYmNCHVpiXDJyA cRJ8gluQtOJR [P].

'Black Holes, Information and the String Theory Revolution', L. Susskind and J. Lindsay (World Scientific 2005) [G].

J. D. Bekenstein 'Information in the Holographic Universe' Scientific American (August 2003) [P]; see also:

> https://ref-sciam.livejournal.com/1190.html [P].

CHAPTER 7: LET'S TWISTOR AGAIN

I obtained the expression in Figure 7.1 from a direct calculation, but the idea for showing this example is from a slide in Bern's public lecture at the Kavli Institute for Theoretical Physics in 2016:

> http://online.kitp.ucsb.edu/plecture/zbern16/options.html [P].

A discussion of MHV diagrams is at:

> https://plus.maths.org/content/picture-perfect [P].

Elvang and Huang's 'Scattering Amplitudes in Gauge Theory and Gravity' (CUP 2014) (see also https://arxiv.org/abs/1308.1697) reviews this area [G].

CHAPTER 8: THE SQUARE ROOT OF GRAVITY

A survey is 'The Double Copy: Gravity from Gluons', C. D. White,

> https://arxiv.org/pdf/1708.07056.pdf [UG].

A graduate-level physics review of this area is 'The Duality Between Color and Kinematics and its Applications', by Z. Bern, J. J. Carrasco, M. Chiodaroli, H. Johansson, and R. Roiban

https://arxiv.org/pdf/1909.01358.pdf

CHAPTER 9: IT'S ONLY PLATONIC

A popular article on the amplituhedron is at:

https://www.quantamagazine.org/physicists-discover-geometry-underlying-particle-physics-20130917/

Arkani-Hamed's 2017 public lecture at the Philosophical Society of Washington describes the background and motivations for some of the later material in this book:

https://www.youtube.com/watch?v=qTx98PUW6lE [P].

A graduate-level summary is Arkani-Hamed, Bourjaily, Cachazo, Goncharov, Postnikov, and Trnka's 'Grassmannian Geometry of Scattering Amplitudes' (CUP 2016) (see also https://arxiv.org/abs/1212.5605) [G].

CHAPTER 10: EVERYTHING IS NOW – NOW

A survey of challenges in the field is Arkani-Hamed's 'The Future of Fundamental Physics', Dædalus 141 (3) Summer 2012 [P]:

https://www.sns.ias.edu/sites/default/files/daed_a_00161(3).pdf

Figures 10.1 and 10.2 were drawn using the Mathematica® package Positroids by J. Bourjaily: https://arxiv.org/abs/1212.6974

CHAPTER 13: STRING THEORY AND REALITY

A detailed talk on the Standard Model is Ellis':

https://www.youtube.com/watch?v=LjR80peV9Os [P, UG].

Index

Printed in the United States
by Baker & Taylor Publisher Services